DISCOVE

STARS & PLANETS

Written by
Toni Eugene

Consultant
Ellis D. Miner

Publications International, Ltd.

Louis Weber, CEO
Publications International, Ltd.
7373 North Cicero Avenue
Lincolnwood, Illinois 60712

Permission is never granted for commercial purposes.

Manufactured in China.

8 7 6 5 4 3 2 1

ISBN: 0-7853-6113-8

Photo credits:

Animals Animals: Roger Archibald: 20; Bruce Davidson: 10; John Lemker: 15, 23; **Getty Images:** FPG International: Front cover, 3, 4, 14, 15, 19, 24; Paul Ambrose Studios: 3, 11; Carmona Photography: 27; Dave Davis: 17; Hansen Planetarium: 10; NASA: Inside front cover, 3, 6, 9, 13, 32, 33, 34; Michael Simpson: 21; Hans Vehrenberg: Inside back cover; Earl Young: 34; Jack Zehrt: 14, 28; **International Stock:** 8, 9, 14, 27, 29, 37; Daily Telegraph Colour Library: 1, 11, 12, 26, 42; **Courtesy of NASA/JPL/Caltech:** 4, 5, 7, 8, 9, 10, 11, 12, 13, 15, 20, 22, 23, 24, 25, 28, 29, 30, 31, 32, 33, 35, 36, 37, 38, 39, 40, 41, 42, 43; **Tom Stack & Associates:** John Cancalosi: 42; Mark Newman: 27; Greg Vaughn: 3, 42.

Illustrations: Pablo Montes O'Neill; Lorie Robare.

Toni Eugene has worked for the National Geographic Society for more than 18 years. She is currently the managing editor in their Special Publications Division. She has authored three children's books, *Creatures of the Woods, Strange Animals of Australia,* and *Hide and Seek.* She received her undergraduate degree in English from Gettysburg College.

Ellis D. Miner, Ph.D., works for the Jet Propulsion Laboratory and has worked on Cassini, Voyager, Viking, and Mariner projects. In 1981 and 1986, he received the *NASA* Medal for Exceptional Scientific Achievement, and in 1990, he was bestowed the *NASA* Medal for Outstanding Leadership. He is the author of *Uranus: The Planet, Rings and Satellites* and *Neptune: The Planet, Rings and Satellites.*

CONTENTS

THE VAST UNIVERSE • 4

THE NINE PLANETS • 20

BEYOND US • 42

GLOSSARY • 44

THE VAST UNIVERSE

is waiting to be explored. Imagine that you could blast off in your spaceship to see it. Long before you reach the stars, you pass the other planets in our solar system that are held in place by our Sun's gravity.

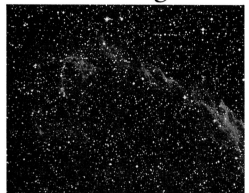

The planets and the billions of stars are millions of miles away. The universe is so vast—so big, so wide, and so deep—it is hard to imagine. If your spaceship traveled as fast as light, it would take 100,000 years to travel across our Milky Way Galaxy. There are many galaxies in our universe.

Every day we learn more about our universe. Come along and discover the stars, planets, galaxies, and other wonders in our universe.

4

GALAXIES

The thousands of twinkling stars that you see in the sky on a clear night are part of our galaxy. If you look closely, you might also see fuzzy glowing patches in the sky. They are masses of interstellar gas and dust. People hundreds of years ago named these glowing patches nebulae, from the Latin word for cloud.

Other glowing patches in the sky are whole systems of stars far out in space. Stars cluster together in systems or groups called galaxies. Galaxies are vast, swirling communities of stars—like neighborhoods of stars.

The stars in a galaxy are held in place by gravity. Everything in a galaxy—all the bits and pieces of rocks and stars—revolves, or moves around, the center, called the nucleus.

The star we know best, the Sun, is part of the community of stars called the Milky Way Galaxy. The Milky Way is the Earth's home galaxy. Scientists once thought the Milky Way was the only galaxy, but now we know that the Milky Way is one of more than 30 galaxies that make up our home cluster, or local group.

Scientists have divided the many billions of galaxies into three types. The Milky Way is shaped like a spiral. It is called a spiral galaxy. A spiral galaxy is shaped a bit like a pinwheel. It has long, curving arms of stars, gas, and dust. From Earth, which is in the Milky Way, we cannot see all of the galaxy.

There are billions of galaxies in the universe, and each galaxy contains from millions to hundreds of billions of stars. Most galaxies, like stars, cluster together in groups. Groups of galaxies are called clusters.

Top: A cluster of galaxies in the constellation Hercules. Center: This spiral galaxy looks much like our home galaxy. The Milky Way Galaxy is also a spiral galaxy.

Left: A nebula is made up of gas and dust. This nebula is called the Trifid Nebula.

What we see of the spiral arms looks like a hazy cloud that arcs across the sky. Some spiral galaxies have tightly wrapped arms, and others have loosely wrapped arms. The center of a spiral galaxy is reddish because old red stars are located there. Most of the light comes from young, bright stars in the disk.

The second type of galaxy is called an elliptical galaxy. Elliptical galaxies do not have arms. They range from nearly round shapes to ovals that are shaped a lot like footballs. Elliptical galaxies seem to have mostly old stars. Unlike spiral galaxies, elliptical galaxies have little gas and dust.

The third type of galaxy is called an irregular galaxy because it has no special shape. It is a formless collection of stars and gas.

One irregular galaxy is called the Large Magellanic Cloud. It looks like an enormous fuzzy swarm of stars. The Large Magellanic Cloud and its near neighbor, the Small Magellanic Cloud, are the closest galaxies to our Milky Way. But they are still far away—it takes a beam of light from the Large Magellanic Cloud 175,000 years to reach Earth. These two irregular galaxies are visible only from the southern hemisphere, the part of Earth that lies below the Equator.

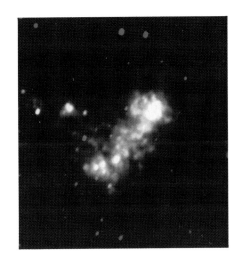

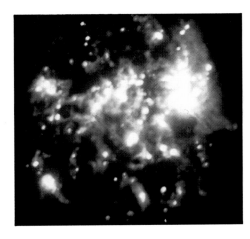

The closest galaxies to our Milky Way Galaxy are the Small Magellanic Cloud (above) and the Large Magellanic Cloud (right).

The billions of galaxies beyond the Milky Way, whatever their shapes, lie in the outer reaches of space—in deep space. Scientists learn more about our galaxy and other neighbor galaxies every year.

A neighbor galaxy to the Milky Way, the Great Galaxy in Andromeda, is a spiral shape. The Andromeda Galaxy is one of the few galaxies we can see without a telescope. It looks like a thin cloud about twice as big as our Moon. But Andromeda is very big and bright. It is brighter than 20 billion suns. Andromeda seems small because it is far away.

THE MILKY WAY

Galaxies are gigantic. It takes a beam of light 100,000 years to travel across the Milky Way. Until 1924, scientists thought the Milky Way was the only galaxy. Now we know that there are billions of galaxies.

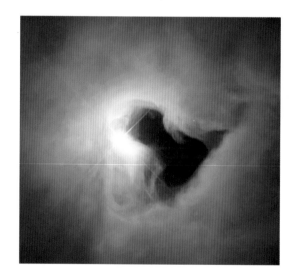

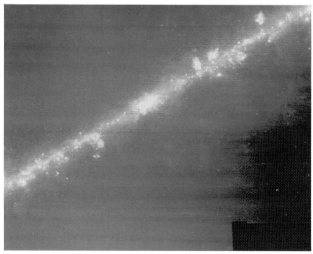

Top: Star systems in the Milky Way. Center: A star forming in the Milky Way. Left: A photograph of the central part of the Milky Way taken by the spacecraft IRAS II.

There are billions of galaxies in the universe, and each contains millions of stars. The Sun, the Earth, the planets, and the stars we usually see in the sky each night are part of a family of stars called the Milky Way Galaxy. For us on Earth, the Milky Way is our home galaxy. Everything in the Milky Way revolves—or moves—around the center of the galaxy. The Sun is our most important star in the Milky Way. The Earth and all the other planets, their moons, and other space rocks revolve around the Sun.

The Milky Way has spiral arms of stars that wrap around it. In the spiral arms of the Milky Way, there are extremely hot blue-white stars. They are giant stars, much hotter than the Sun. The center of the Milky Way Galaxy, where the four parts of a pinwheel join together, glows orange red. That glow comes from red giant stars—old stars that are not as hot or as bright as they were billions of years ago. Within the pinwheel of the Milky Way are deep red patches. These glowing patches are called nebulae. They are vast areas of gas and dust where new stars are born.

Huge collections of stars form balls like loosely packed snowballs that surround the center of the Milky Way. These huge balls of stars are called globular clusters. There are more than 100,000 different globular

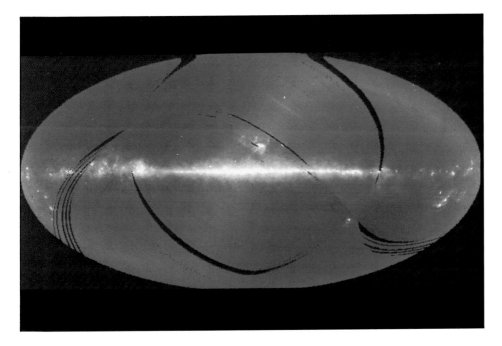

The Milky Way is only a small part of the universe. And our entire solar system—the Sun and all the planets—is just a small part of the Milky Way.

clusters surrounding the center of the Milky Way. Each cluster contains more than 100,000 stars. In the Milky Way there are billions of bigger and brighter stars than our Sun, and billions of stars smaller and dimmer. The Sun is important to us on Earth because it is the center of our solar system.

Above: A full view of the Milky Way Galaxy taken by the spacecraft IRAS.

The Sun and the planets of our solar system, including Earth, are in a large spiral arm of the Milky Way. That arm of the galaxy is called the Orion Arm. Our solar system is about two-thirds of the way from the center of the Milky Way in the Orion Arm. Our Sun, with its family of planets, constantly moves around the center of the Milky Way. It moves very fast, but the Milky Way is so big that from Earth we don't even notice that we are moving. The Milky Way is so big that it takes our solar system 230 million years to circle around it one time.

Above and left: Because our solar system is part of the Milky Way, we cannot see what the Milky Way looks like. We are too close to see all of our home galaxy. What we do see of the Milky Way from Earth is a hazy band of clouds that arc across the sky.

THE SOLAR SYSTEM

Just as your family is made up of separate people who are related, the solar system is made up of different objects that are related to the Sun.

Above: Earth is the only planet in our solar system that we know has life on it. Left: The Sun is the center of our solar system.

Earth is a planet that moves around the Sun. The Sun is a star. Its light gives us day, and when that light is not shining on our part of the Earth, we say it is night. The Sun makes life here possible. Without the Sun, it would be too cold to live on planet Earth.

The Earth is only one of many objects held in place by the Sun's gravity. Other objects revolve around the Sun, too. All these objects make up the solar system. The word "solar" means Sun. The solar system is the family of objects that revolve around the Sun.

The Sun has 99 percent of all the material in our solar system. It is the biggest object in our solar system. Including the Earth, nine planets are in the solar system.

The planet nearest the Sun is Mercury. Then come Venus, Earth, and Mars. Beyond these four inner planets are a band of large chunks of rock and metal called asteroids.

Tens of thousands of asteroids revolve around the Sun beyond Mars. This band of rock and metal chunks is called the asteroid belt.

Sometimes asteroids ram into each other in space and break into smaller pieces. Comets also break up, and the pieces are called meteoroids. Meteoroids sometimes are near Earth and appear as bright streaks in the night sky. The meteoroids that get close enough to glow in our sky are called meteors. The few meteors that reach the surface of Earth are called meteorites.

Beyond the belt of asteroids lies the biggest planet in the solar system—Jupiter. Jupiter is a ball of gases. So are the next planets: Saturn, Uranus, and Neptune. They are called the gas giants.

The farthest planet from the Sun is Pluto. It is the smallest known planet and is different from the other four outer planets. Pluto is an icy ball smaller than Earth's Moon.

Beyond Pluto, at the edge of our solar system, comets also revolve around the Sun. Comets are ice and dust. Millions of comets circulate the Sun, but they can be seen only when they get close enough to the Sun that its heat makes them glow. The gravity of other comets and planets can change their path so that they come into the inner portions of the solar system, where we can see them. Eventually, comets evaporate and become grains of dust.

Scientists think our solar system formed about 4½ billion years ago. A huge swirling cloud of dust and gas—a nebula—grew smaller and smaller. Over millions of years, the innermost pieces drifted toward the center and became hot and formed our Sun. The more distant parts of the cloud contained what would become the rest of the solar system.

Jupiter is one of the gas giants.

All the planets, their moons, asteroids, and comets revolve around the Sun. They are all a part of our fascinating solar family called our solar system.

Above: Comets are held in our solar system by the Sun's gravity. Left: Scientists are constantly trying to learn more about our solar system. They have sent spacecraft to both comets and asteroids.

EARTH'S SUN

Right: The Sun is so big that it holds objects in our solar system around it by a force called gravity. Earth, the other eight planets, asteroids, and comets move around the Sun because of its gravity. Below: During a solar eclipse, the corona of the Sun is visible. Usually, the brightness of the Sun makes the corona invisible.

The Sun is very powerful. It makes our daylight, makes our weather, and makes our crops grow. Without the Sun, no life would be on Earth!

The Sun's symbol

Every day, whether it is a bright, hot day or a cold, rainy day, the Sun shines light on Earth. Sometimes daylight is bright. Sometimes clouds hide the light, but there is still light from the Sun, which is a star.

Of billions of stars in the Milky Way Galaxy, the Sun is the star closest to Earth. It is the star we know best. A lot of what we know about stars, we learned from studying the Sun. The Sun is 93 million miles away from Earth. A beam of light from the Sun travels that distance in eight minutes and 20 seconds.

The Sun is by far the largest thing in our solar system. Everything else in the solar system is very small compared to the Sun. And of all the stars in the universe, the Sun is only a medium-size star.

The Sun formed about 4½ billion years ago. Scientists believe it formed from a whirling cloud of dust and gases. Like all stars, the Sun spins—much as a top spins when you twirl it, and the Sun, like all stars, is always changing. Nuclear explosions occur constantly inside the Sun.

The Sun is made mostly of a gas called hydrogen. Scientists have found that the Sun is formed of four layers. The outer layer is called the corona. It glows. Usually, the Sun is so bright that we can't see the corona. If you can watch the Sun when the moon is passing in front of it, you can see the glow of the corona.

The next layer of the Sun is the lower atmosphere. It is also called the chromosphere. The chromosphere is always exploding. Violent eruptions like huge volcanoes occur all the time. Some of the largest of these explosions are called solar flares.

The surface of the Sun, the part we see most easily from Earth, is called the photosphere. The fourth—and last—layer of the Sun is called the core. It is the middle of the Sun, where all the heat and light we feel and see on Earth is produced.

The heat and light that the Sun produces are two kinds of energy. Both the heat and light come out of—or radiate from—the Sun. This heat and light radiation travels in waves. Just as the waves in the ocean come in different sizes, radiation from the Sun travels in waves of different sizes.

The only light we can see travels in wavelengths that make up a rainbow. A rainbow has bands of color —red, orange, yellow, green, blue, and violet. These are the colors our eyes can see.

Other forms of light—or radiation —reach Earth from the Sun. We cannot see these forms of light. We feel infrared rays as heat. When you sit out on the beach and get hotter and hotter, infrared rays from the Sun are making you hot. That same day at the beach, you will get either burned or tanned. The rays that tan or burn you are called ultraviolet rays.

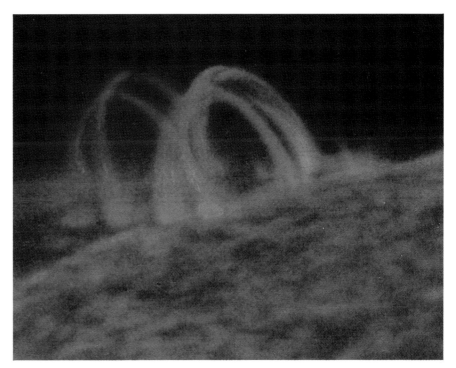

Skylab took this picture of a solar flare—a giant loop of hydrogen gas that erupts. Some solar flares can cause magnetic and electrical problems on Earth.

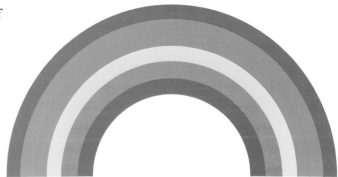

The Ulysses spacecraft continues to study the Sun and interstellar space.

The Sun gives off energy that can make a rainbow, make you hot, and tan your skin.

13

STARS

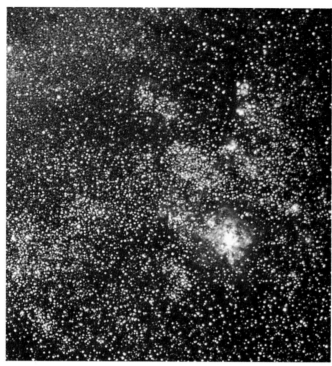

Stars are different colors. Old stars are redder than younger stars, which are white or blue. Can you spot an older star in this star system?

A star is like a ball of fire. The outer gases of a star are so hot that they bubble like a boiling pot. Below the surface, it is even hotter.

Right: Nebulae, such as the Horsehead Nebula in Orion, are where stars are born.

At the middle of a star—the core—temperatures and pressure are so great that particles ram each other so hard that they stick together. This process is called fusion, which produces energy in the form of light. Fusion makes a star shine.

Stars are different colors. The hottest stars shine with a blue-white light. They are young stars. Some stars shine with a red light. Red stars are cooler than blue stars and are usually older. Some stars seem to glow yellow or orange—they are in the middle of their lives. The color of a star depends on how hot it is, how big it is, and how much energy it produces.

The rate at which a star produces energy is called luminosity. Stars are different sizes, and some have more gases than others. The more massive a star is, the more luminous it is. Each star has a different brightness. How bright a star appears from Earth is called apparent brightness or apparent magnitude. But stars are not all the same distance away, so it is difficult to compare their real brightness.

Scientists rate stars as if they were the same distance from Earth. That measurement of brightness is called absolute brightness or absolute magnitude. The absolute magnitude of a star far from Earth is greater than its apparent magnitude.

Stars are formed—or born—in nebulae. A globe of matter called a protostar grows in the spinning cloud of dust and gases. The gravity

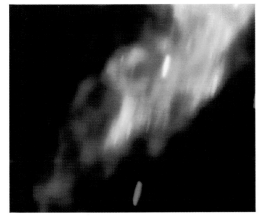

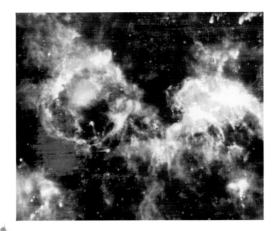

This is the beginning of two new stars!

of the protostar attracts more matter, and the protostar grows. As it gets bigger, it grows hotter and brighter. Its gravity continues to draw more material, which gets packed tighter and tighter. The protostar core heats up and glows red. It gets so hot and pressure increases so much that fusion occurs, and the star begins to shine, giving off energy as light.

After millions—or even billions—of years, a star begins to burn its heavier gases. Over time, it sends out less energy and begins to cool. Less massive stars shine longer than more massive stars, but all stars will die.

Some stars are so special that they stand out. Supergiants are very large, massive stars that are really bright. One of the most famous supergiants is called Betelgeuse. It is about 520 light years from Earth. Betelgeuse is a red supergiant. Betelgeuse is very bright, but it is cooler than some stars because its surface is so gigantic that lots of energy can be released all the time. Betelgeuse is about 800 times bigger than our Sun. Rigel is a star 900 light years away and shines so brightly we can see it even without telescopes. Rigel is a blue supergiant.

Right: A ring nebula. Below: Both Betelgeuse and Rigel are part of the constellation Orion at the bottom of the picture. The three evenly spaced stars that are between Betelgeuse and Rigel are called Orion's belt.

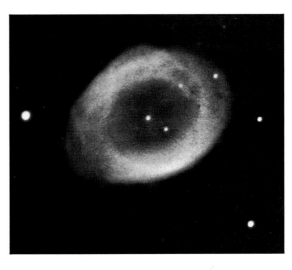

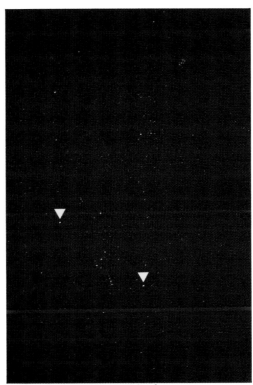

Betelgeuse   Rigel

Tonight, look out your window at home. Can you spot Betelgeuse or other famous stars? Stars are one of the many wonders of our universe.

15

**LIBRA
THE SCALES**

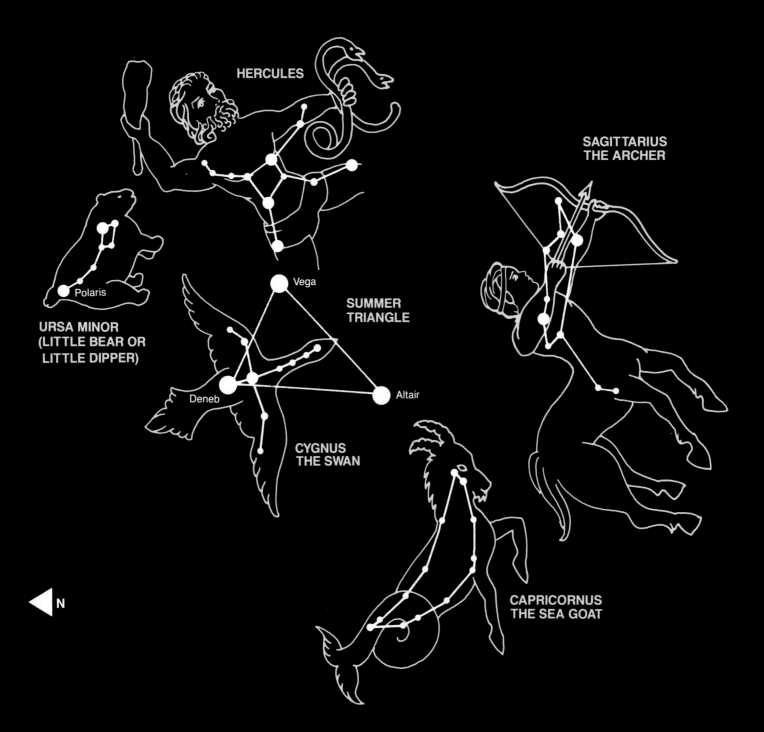

HERCULES

**SAGITTARIUS
THE ARCHER**

Vega

**SUMMER
TRIANGLE**

Polaris

**URSA MINOR
(LITTLE BEAR OR
LITTLE DIPPER)**

Deneb

Altair

**CYGNUS
THE SWAN**

N

**CAPRICORNUS
THE SEA GOAT**

CONSTELLATIONS

When you look into a sky full of clouds, you can sometimes make imaginary shapes out of them. One cloud might look like a sailing ship, and another like a puppy. For thousands of years, people have thought up imaginary shapes made of stars, called constellations.

Historians think the first constellations were the 12 signs of the Zodiac. The Zodiac is an imaginary path of stars the Moon and the planets appear to travel. As the Earth moves around the Sun, the slice of sky we see changes. It takes a whole year for a full circle of sky to pass. The stars are like a twinkling curtain. The Earth and all the objects in the solar system move in front of that curtain of stars.

When you tell people your birthday, someone may say, "Oh, you're an Aries," or, "You must be a Leo." That is because each year is divided into 12 constellations. And each constellation, if you really use your imagination, has a shape. That shape is its sign. Leo is a lion. Aries is a ram. Do you know what your constellation is?

Now people who study stars say there are 88 constellations in the sky. There are land animals, birds, water animals, people, and even a dragon and a unicorn. Each constellation has a Latin name.

Opposite page: To see a night sky with these constellations in this order, you must be looking at a late summer sky.

Left: The symbols of the 12 signs of the Zodiac, from top, just left of center and going counterclockwise: Aries, Taurus, Gemini, Cancer, Leo, Virgo, Libra, Scorpio, Sagittarius, Capricorn, Aquarius, and Pisces. Below: The stars are constantly moving. This chart shows the Big Dipper's path around Polaris.

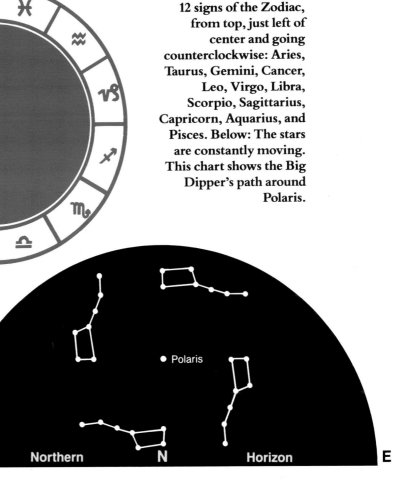

W   **Northern**   **N**   **Horizon**   E

CONSTELLATIONS

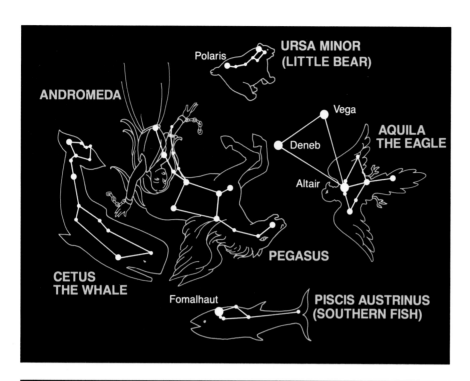

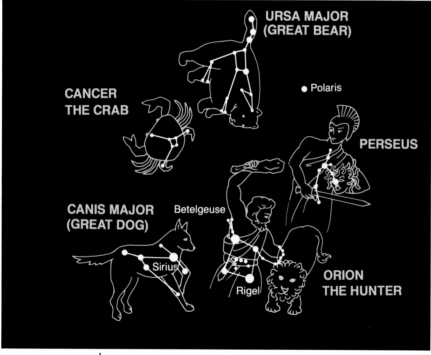

Top: This is a late fall sky. Center: Look at a late winter night sky to see these constellations in this order.

The Earth is so big that we cannot see all the stars at the same time. The Earth is round, so parts of the sky are always out of sight beyond the horizon. If you lived at the very middle of the Earth—at the Equator—you could see all the stars as they passed. In a year, all the stars would pass by you. But if you don't live at the Equator, you can see only part of the sky. The area above the Equator is called the northern hemisphere. Below the Equator is the southern hemisphere.

One of the most famous constellations in the southern hemisphere is the Southern Cross. You can see it only in the southern hemisphere. The Crux, or Southern Cross, is made of four stars that look like a small cross.

In the northern hemisphere you can see the Northern Cross. It, too, looks like a big cross. It is part of a constellation called Cygnus the Swan.

One of the easiest shapes to see in the northern hemisphere is the Big Dipper. It looks like a big pot with a long handle. The Big Dipper is made of seven bright stars. Those stars are part of a constellation called Ursa Major, or the Great Bear.

Another constellation to look for in the northern hemisphere is Ursa Minor, or the Little Bear. Seven stars in Ursa Minor make the Little Dipper, another shape that is easy to find. Like the Big Dipper, the Little Dipper looks like a big pot. The star at the very tip of its handle is Polaris, or the North Star. Sailors navigated by Polaris hundreds of years ago. And sailors and astronauts still use it today.

Most constellations are not as easy to see as the Big and Little Dippers. You have to really look and use your imagination to see some of the constellations. But learning about constellations and locating them in the sky is a great way to learn more about the stars themselves. If you can find a particular constellation among all the stars in the sky, then you can focus on the stars in that constellation.

Orion is a constellation you can find in the evening winter sky or morning summer sky in the northern hemisphere. From December to March, Orion is the main evening constellation. Orion the hunter holds a lion in his left hand and a club in his right hand. A sword hangs from his belt. Orion contains more bright stars than any other constellation. The blue supergiant Rigel and the red supergiant Betelgeuse are both in Orion. So is another very bright star called Bellatrix.

Because everything in the galaxy is moving around the center of the galaxy, the stars are always moving. So the constellations are always changing shape. The stars are so many light-years away that they don't seem to be moving.

Even today, astronauts and other travelers use Polaris to find out where they are.

You must use your imagination to see the constellation shapes that ancient people saw in the stars.

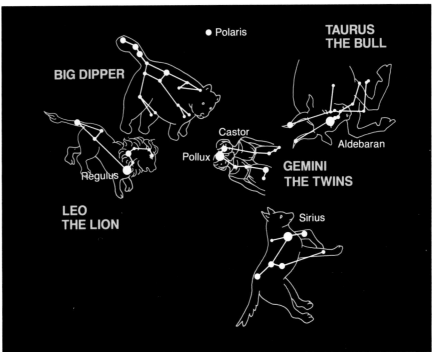

Above: You will be able to see these constellations in the evening sky in early spring. Right: If life exists in other galaxies, do you think the people group stars together in shapes and name them?

THE NINE PLANETS

of our solar system are waiting to be explored in your imaginary spaceship. These planets are held in place by the gravity of our Sun—the center of our solar system.

The path of a planet around the Sun is called its orbit. On each planet, you might want to know how long it would take until your next birthday. As a planet moves around the Sun, it is revolving. Every revolution is a year on that planet.

We also need to know how long a day is on each planet. Each planet spins, or rotates, around an imaginary line that runs from the south pole to the north pole—called its axis. One rotation equals one day on that planet.

MERCURY

Mercury is one of the hottest and coldest planets in our solar system!

Mariner 10 took this picture of Mercury's rocky, cratered surface.

Mercury was named after a Roman god, who was the messenger of the gods.

Mercury, the planet closest to the Sun, is also the fastest planet in the solar system. Mercury revolves around the Sun in only 88 days compared to Earth's 365-day year.

Mercury revolves quickly, but its rotation is very slow. It takes the planet 59 Earth days to complete one rotation. Because of the slow rotation and the fast revolution around the Sun, sunrises occur only once every 176 Earth days.

The long days on Mercury and its nearness to the Sun help make it one of the coldest and hottest planets in the solar system. The side of Mercury facing the Sun bakes in daylight at more than 800°F. On the night side of the planet, temperatures drop to minus 274°F. No other planet has such a wide temperature range.

Mercury is a bare and rocky ball covered with deep holes, called craters. The craters formed when meteorites crashed into Mercury billions of years ago. The biggest crater is called Caloris. It is 800 miles across —bigger than the whole state of Texas.

Mercury has no satellites of its own. It is a world with almost no atmosphere, or air. There are no rivers and no oceans. Almost nothing has changed on the planet since shortly after the solar system formed billions of years ago.

Mariner 10 traveled nearly 100 million miles through space before it got close to Mercury. The trip took 146 days. Mariner 10 passed close to Mercury three times. Each time it photographed and mapped the planet. Television cameras on the spacecraft sent the pictures through space back to Earth.

Mariner 10 discovered that huge, steep cliffs up to two miles high cut through Mercury's rocky surface. The cliffs slice right across the walls of craters and are hundreds of miles long. Lava from old volcanoes has created wide plains on one side of the planet, and the entire surface of Mercury is covered with a thick layer of dust. Scientists have known for a long time that Mercury is very dense —that the material it is made of is tightly packed together so it weighs a lot. (A rock the size of your lunchbox weighs a lot more than your empty lunchbox. The rock is denser.) Mariner 10 proved that Mercury has a very large core of iron, which makes it very dense. It is almost as dense as Earth, but contains less than a tenth of the material of Earth.

About 14 times every hundred years, Mercury passes right between the Earth and the Sun. It moves from east to west. Scientists call this crossing a solar transit. Then Mercury looks like a small black dot against the bright circle of the Sun.

On this January morning in 1984, four planets could be seen in the sky. The largest glowing object is Venus. Below and to the left, just above the orange sunrise, is Jupiter. If you look very closely, to the left of Jupiter is a small white object, that is Mercury. Antares, a red star, is straight to the right of Venus. To the top right of the picture is Saturn.

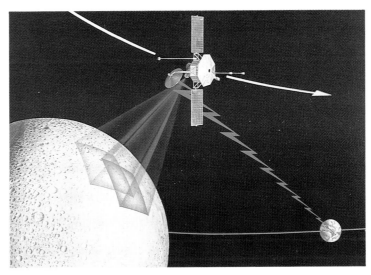

Mariner 10 was the first spacecraft to explore Mercury.

Mercury was fast because of his winged sandals. The planet is also quick; its planetary year is 88 Earth days!

VENUS

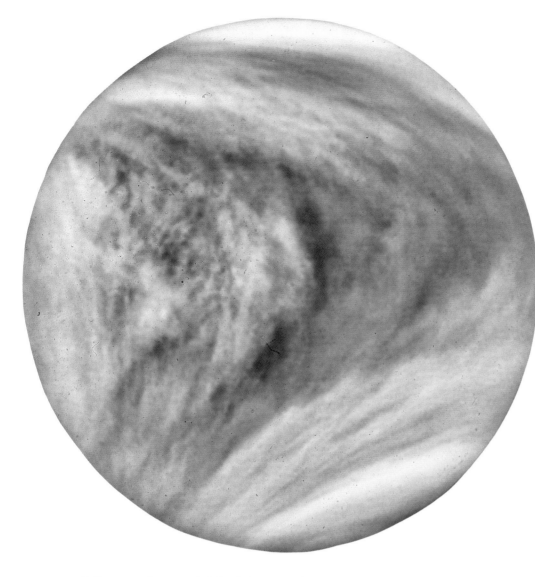

More than 100,000 volcanoes dot the surface of Venus, the second planet from the Sun.

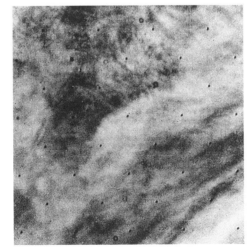

Venus was named for the Roman goddess of love and beauty. The sign of the planet is the hand mirror the goddess used to admire herself.

Venus is unbelievably hot, but it is also cloudy. Only a little light gets through the deep clouds that cover the planet. There is no water and no oxygen. Life on Venus is impossible for humans.

Not long ago, people thought Venus was like Earth. They called it Earth's sister planet. It is like Earth in size, mass, and density. But it is also very different.

The orbit of Venus is nearly circular. Because it is twice as big as the orbit of Mercury, scientists can see Venus outside of the Sun's glare. Venus completes one revolution in 225 Earth days.

The rotation of Venus is the longest in the solar system. It takes Venus 243 Earth days to complete one rotation. Venus also rotates backward. Earth rotates the same direction as its revolution around the Sun. Venus rotates opposite to its revolution around the Sun. Scientists call this retrograde—backward—rotation. Meanwhile, the time between sunrises on Venus is 117 days!

The thick clouds that cover Venus make it bright. They reflect the light of the Sun. Those same clouds make it hard to study Venus. They are so thick that they completely hide the surface. Scientists finally received a good look of Venus when spacecraft bounced radar off the surface and sent back the information to Earth.

Mariner 10 took this close-up picture of the clouds covering the surface of Venus.

In 1974, Mariner 10 found that Venus was very dry and sizzling hot. Pioneer Venus 2 sent back information about three blankets of clouds around the planet. The upper layer is poisonous. The clouds are made of sulfuric acid, and they swirl at almost 200 miles an hour. The thin, hazy upper layer of clouds is 15 miles thick. Below it, another layer of clouds 40 miles thick covers the surface of the planet. The atmosphere at the surface of Venus is almost all carbon dioxide. (Humans breathe in oxygen and breathe out carbon dioxide.) The atmosphere of carbon dioxide is very heavy.

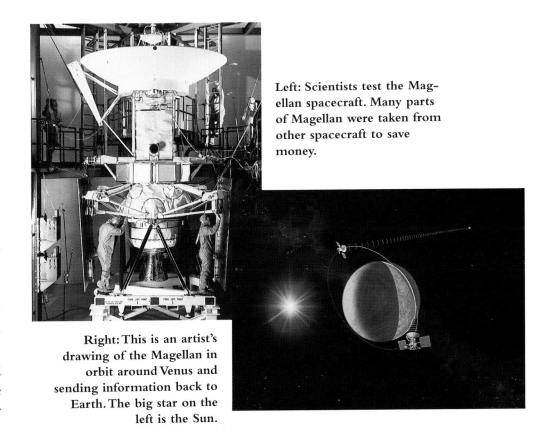

Left: Scientists test the Magellan spacecraft. Many parts of Magellan were taken from other spacecraft to save money.

Right: This is an artist's drawing of the Magellan in orbit around Venus and sending information back to Earth. The big star on the left is the Sun.

The thick atmosphere of carbon dioxide on Venus lets in radiation from the Sun, which turns to heat. When the heat hits the ground, it bounces back into the atmosphere, just as a hot street sends off heat on Earth. But the heat cannot escape. On Venus, the heavy atmosphere traps the heat. The heat stays on Venus, and temperatures reach 900°F.

The United States space probe Magellan began orbiting Venus in 1990. It carried radar that could go through the clouds. Magellan radar images covered 99 percent of the planet.

The Magellan spacecraft took this picture of Venus. It used radar to strip the planet of its clouds so we could see the surface of the planet.

Earth changes all the time. Scientists say that it is a "living" planet. Over time, rain and other weather wear away even mountains.

From space, Earth looks like a huge blue ball with white streaks across it. Earth gets its blue color from the oceans that cover much of its surface and from the air that surrounds it. The white streaks are clouds. Earth is the only planet in the solar system that we know has life on it.

EARTH

Earth is the third planet from the Sun. Like Mercury and Venus, it is a solid ball.

Earth's orbit is slightly oval. The planet rotates on its axis in 23.9 hours. However, because the Earth also revolves around the Sun while it spins, the time between sunrises is 24.0 hours—our day. In that time, one side of the globe is in daylight, and the other side is in darkness. Earth completes one revolution around the Sun every 365.25 days.

Earth is tilted on its axis. The Earth's tilt causes the seasons. When the top of the Earth—the North Pole—tilts toward the Sun, it is summer in the northern hemisphere. When the North Pole tilts away, it is winter. Winter in the northern hemisphere is summer in the southern hemisphere.

The atmosphere of Earth is made mostly of two gases—nitrogen and oxygen. These gases make it possible for life—humans breathe oxygen.

Earth's atmosphere lets heat in and out, and it is layered. Most of our atmosphere, from the surface of the planet going up, is six miles thick. A very important layer of our atmosphere is the ozone layer. It shields us and filters out the Sun's harmful rays. Without the ozone layer, life on Earth would die. That is why scientists and other people are concerned that we don't put harmful chemicals in the air that might destroy the ozone layer.

Water is the key to life on Earth. Oceans cover nearly three quarters of the planet's surface.

The planet was named by the ancient Greeks for their goddess Gaea—Mother Earth. The symbol of Earth is the Greek symbol for sphere or ball.

Left: This ultraviolet photograph of Earth shows where the Sun's heat is hitting the planet. It is daytime in most of the red spots and nighttime on the other side of the planet.

Right: Earth's weather keeps changing what our planet's surface looks like.

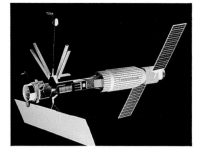

The spacecraft Seasat (right) allowed scientists to see what the ocean floor looks like (below).

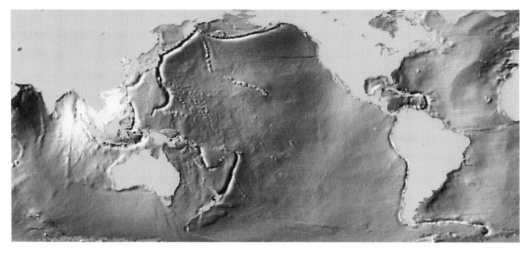

Some of the most exciting formations on Earth are hidden. They are on the ocean floor. Water covers the deepest canyons—called trenches—and some of the tallest mountains.

This is the beautiful view astronauts get of the Earth and the Moon from their spaceship window.

The oceans store heat and keep temperatures on Earth comfortable. Water evaporates from the surface of the oceans—the heat from the Sun turns water into a gas or vapor. The water vapor falls back to the Earth as rain. Earth is the only known planet with oceans of liquid water. Most living things on Earth—including you—are made mostly of water.

When Earth formed, it did not have oceans. More than four billion years ago, Earth was very hot. Gases erupted from volcanoes. Millions of years passed. The gas became clouds and turned to rain. Over time, that rain formed our oceans. Water made life on Earth possible.

Unlike Mercury, Earth looks very different from the way it used to. Volcanoes, like Kilauea on the island of Hawaii, erupt, building new land. Earthquakes shake the land.

The crust of Earth covers the planet, much as skin covers your body. The crust is a thin layer of rock with craters, mountains, and rivers. Below the crust is another layer of rock, called the mantle. The part of the mantle deep below the crust is melted—or molten—rock. The core is fluid rock around a solid center.

Much of the Earth is hidden. The Mariana Trench on the floor of the Pacific Ocean is 36,000 feet—more than seven miles—below the surface. Some ocean mountains are so high that they form islands. Iceland is the top of an underwater mountain!

THE MOON

The Moon is Earth's satellite. It revolves around the Earth and is held in place by Earth's gravity. The Moon is about one quarter the diameter of Earth. It is about 240,000 miles away.

The Moon seems to change—sometimes we see only a sliver. Sometimes we see a great round yellow circle. That is because the Moon has days and nights, just as the Earth does. At times, shadows hide part of the Moon.

The Moon revolves around the Earth every 27.3 days. The Moon is frozen in its rotation—it completes one rotation in the same time as it completes one revolution. Because of this, we always see the same side.

In 1969, the first people landed on the Moon. Their spacecraft, the Apollo 11, was launched by NASA (National Aeronautics and Space Administration). Neil Armstrong was the first person to walk on the Moon! The Moon's gravity is much less than Earth's—so he bounced rather than walked.

The Moon has huge basins that were made when asteroids rammed it. Then, lava filled the basins. Astronomers call them maria—the Latin word for seas. One basin, called the Sea of Tranquillity, is 600 miles across. It is where Neil Armstrong landed, and his footprints are still there!

The Moon is a dry, dead place. There is no atmosphere and no water. It has not changed since shortly after the solar system formed.

The moon was named Diana or Luna by the ancient Romans. Luna means "shines." The symbol of the Moon is a crescent moon.

Left: Neil Armstrong took this picture of fellow astronaut Buzz Aldrin on the Moon.

MARS

Mars was named after the Roman god of war.
The symbol of the planet stands for the god's shield and spear.

For hundreds of years, scientists have wondered about life on Mars, the fourth planet from the Sun. Stories and movies have been written about "martians"—aliens from Mars. The stories are fun, but scientists now know that there is no life on Mars.

The orbit of Mars is much larger and slower than Earth's. It takes Mars 687 Earth days to revolve around the Sun. And Mars's orbit is off center. Sometimes the planet is much closer to the Sun than other times. Mars rotates on its axis every 24 hours and 37 minutes. So a day on Mars is just a little longer than a day on Earth.

Mars ●   ⬤ **Earth**

Mariner 9 (left) sent back photographs of Olympus Mons (above)—the biggest volcano we know. It is 15 miles high, has a crater 45 miles wide, and is 355 miles wide at its base. Olympus Mons is 20 times bigger than the biggest volcano on Earth!

Like Earth, Mars has a rocky core. But the planet is only about half the diameter of Earth. Because it is much farther from the Sun than Earth, Mars gets half as much energy from the Sun as Earth does. Because Mars has a very thin atmosphere, much of the Sun's energy, which heats the surface, escapes back into space. It is very cold on Mars. On a hot summer day, temperatures might rise above freezing, but usually they are far below zero.

**Mars is a frozen world that is covered with red dust.
The orange-red dust that covers the surface of the planet earned Mars its nickname—the Red Planet.**

Two small moons revolve around Mars—probably pieces of asteroids. They are irregular-shaped blobs covered with craters. The moons are named Phobos and Deimos—Fear and Terror—after the two sons of Ares, the Greek god of war.

Since 1965, a variety of spacecraft, which includes Mariners 4, 6, 7, and 9, Vikings 1 and 2, Pathfinder, and Global Surveyor, have visited Mars. Together, they have shown us many different landscapes of the planet, such as enormous volcanoes, craters formed when meteoroids smashed into its surface, the largest canyon in the solar system, and channels that appear to have been carved by flowing water. You can read more about these findings on the Internet at *http://mars.jpl.nasa.gov/*.

These missions, the Mars Odyssey (launched in 2001), and other spacecraft planned for Mars are intended to prepare the way for a manned mission to Mars sometime in the 21st century. Mars appears to be more like Earth than any other planet in our Solar System, and many scientists believe we may find traces of past microscopic life in the soil of Mars. Even if no evidence for life on Mars is found, the absence of life may help scientists understand how life on Earth began.

Left: Viking 1 and 2 were sent to Mars to further explore the "Red Planet." Below left: Both Viking 1 and 2 sent landers to land on Mars for a closer look. Below: Mars Global Surveyor was launched November 7, 1996, and entered orbit on September 12, 1997.

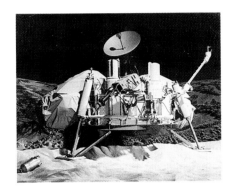

Some scientists want to send astronauts to Mars to establish a space station there. They would like to do that within the next 30 years—in your lifetime!

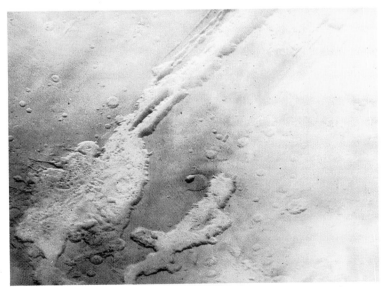

Mariner 9 found a gigantic valley on Mars that makes a deep scar across the middle of the planet. The valley was named after the Mariner spacecraft—Valles Marinaris. It is more than ten times as long as our Grand Canyon and three times as deep —more than three miles from the top to the bottom.

31

JUPITER

A voyage from Mars to the fifth planet from the Sun—Jupiter—could be very dangerous. Your spaceship would have to pass through the asteroid belt—millions of miles occupied by thousands of asteroids. The chances of being hit by an asteroid are small. But if one hit your spaceship, you and your spaceship would not survive.

Jupiter is five times farther from the Sun than Earth. Jupiter is the first of the outer planets—or gas giants. It is the biggest planet in the solar system. Unlike Earth and the other inner planets, Jupiter is made mostly of gases, but it probably has a molten rocky (liquid rock) core.

It takes Jupiter 11.9 Earth years to revolve around the Sun. But Jupiter has the shortest day in the solar system. It completes one rotation in only 9.9 hours. The planet's swift rotation makes the clouds swirl on the planet's surface.

Jupiter, like the Sun, is made almost entirely of hydrogen and helium gases. If the planet were very much bigger, it might have been a star. Because it is still cooling from the time of its formation, it gives off nearly twice as much heat as it receives from the Sun.

Spacecraft that flew by Jupiter sent scientists information they could not get from watching the planet through telescopes. Voyager 1 was launched in 1977 and approached Jupiter in 1979. It measured Jupiter's atmosphere and found that it is thousands of miles thick.

Jupiter is two and a half times bigger than all the other planets combined. It is as big as 1,300 Earths.

Jupiter is named after the king of the Roman gods. Its symbol is the sign for the lightning bolt.

Jupiter's rings are beautiful when lit up by the Sun.

Jupiter has 39 moons. Four of those moons—the biggest ones—were discovered by a man named Galileo nearly 400 years ago. They were the first new objects in space discovered by looking through a telescope. Jupiter's four big moons are called the Galilean moons after the man who discovered them. They are Io, Europa, Ganymede, and Callisto.

Left: Jupiter's Great Red Spot is a giant storm in its atmosphere. It is so big that three Earths could fit in it! Below: Jupiter is shown with four of its 39 moons.

Jupiter is a world of swirling bands of colored clouds, which move at speeds up to 260 miles per hour. The atmosphere of Jupiter is like a bubbling pot of paints that swirl but don't mix. Near the cloud tops, lightning is always flashing.

As scientists at the Jet Propulsion Laboratory in California studied Voyager pictures of Io, they became very excited. They found a volcano erupting on Io (see above). Until that moment, we thought only Earth had active volcanoes. Voyager showed us that Io, unlike most other objects in the universe, is very young. It is changing all the time.

Spacecrafts, such as the Voyagers, have helped astronomers learn a lot about the planets. In 1989, NASA launched a spacecraft that reached Jupiter in 1995. They named the spacecraft Galileo. Galileo dropped a probe into Jupiter's atmosphere of swirling clouds, and scientists learned much more about the biggest planet in our solar system.

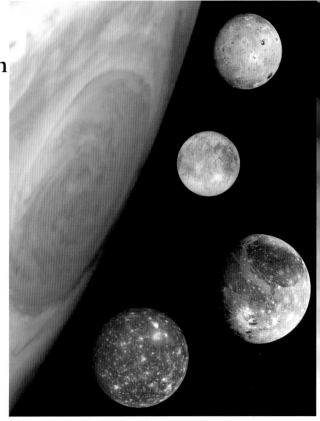

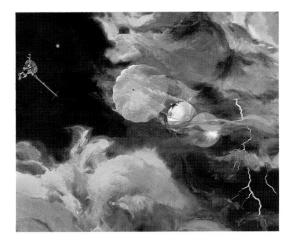

Left: This is an artist's drawing of the Galileo-Probe entering Jupiter's atmosphere.

33

SATURN

Saturn, the sixth planet from the Sun, is the second largest in the solar system. Saturn is the most distant planet you can see easily without a telescope. Until about 200 years ago, scientists thought Saturn was the last planet in our solar system.

Like Jupiter, Saturn is a gas giant. It is a cold world that receives only one hundredth of the heat and light from the Sun that Earth receives. Temperatures in Saturn's atmosphere drop to 290°F below zero.

It takes Saturn more than 29 years to revolve around the Sun. Imagine—you would have to wait 29 Earth years to have a birthday on Saturn. But Saturn spins rapidly. It rotates on its axis once every 10.7 hours. Saturn spins so fast that the big gas giant bulges in the middle.

Galileo discovered the rings around Saturn. To him they looked like ears, and for a long time Saturn was called "the eared planet." Scientists once thought Saturn had seven rings. Thanks to the Voyager spacecraft, we know much more now. We know that Saturn has hundreds of rings around it that extend for thousands of miles. The rings are probably pieces of ice that are from pebble-size to house-size. The rings are very thin—only about 40 feet thick.

Saturn looks like a huge round blob of butterscotch pudding. Haze and rings surround the planet. It is very beautiful.

The planet is named after the Roman god of the harvest. The symbol of the planet is a sickle, the curved blade on a pole people once used to cut down their crops.

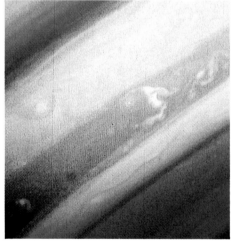

Saturn's atmosphere is mostly hydrogen and helium. The atmosphere of Saturn has belts and zones, like Jupiter's. And the atmosphere moves fast —at speeds up to 1,100 miles per hour.

Enceladus

Tethys

Dione

Enceladus, a primary moon of Saturn, is made of ice. It reflects light like freshly fallen snow does. Tethys and Dione are also made of ice. The other four primary moons of Saturn are Mimas, Rhea, Iapetus, and Hyperion. Hyperion looks like a potato.

Saturn has 30 moons, of which 13 have irregular orbits. Most of Saturn's moons are balls of ice. Eight of them are middle-size to big. They are called primary moons. Titan, the largest, is the only satellite in the solar system that has an atmosphere similar to Earth's. Temperatures on Titan drop to almost 300°F below zero—much colder than any place on Earth.

Pioneer and Voyager spacecrafts launched by NASA sent scientists a lot of information about Saturn. NASA is still studying the photographs. Until August 1990, scientists knew only 17 moons orbiting Saturn. Scientists studying Voyager 2 photographs found an 18th moon. It is one of the smaller moons of Saturn, and its gravity keeps the ice particles in one of the planet's rings. When a moon does this, it is called a shepherding moon. The moon is named Pan, after the Greek god who tended sheep. Twelve more moons were found in 2000.

NASA launched the spacecraft Cassini in 1997; it will reach the planet in 2004. Scientists at NASA hope to discover more surprises about the butterscotch-colored planet with rings.

Scientists hope that the Cassini spacecraft will uncover new and interesting things about the butterscotch-colored planet.

Above: Seven of Saturn's rings are named. They are the A Ring, B Ring, and C, D, E, F, and G Rings. The A Ring is the outermost ring we can see from Earth. It is so big that it would take you 95 years to walk around it!

URANUS

Uranus, the seventh planet from the Sun, is only a faint gleam in the night sky. Uranus is more than one and a half billion miles from the Sun! Even with a telescope, it is very hard to see Uranus.

Uranus was named after the Roman god of the heavens. The symbol of the planet is the sign for the metal platinum.

Scientists knew that Uranus had nine rings, but Voyager 2 discovered a tenth! The rings are only a few miles wide. They look like black hoops. Scientists think the rings are made of darkened methane ice.

It is very dark and cold on Uranus. The Sun is only a tiny circle of light, and temperatures in the atmosphere are about 350°F below zero. A music teacher who studied the sky for fun discovered Uranus in 1781, with a telescope he built himself.

Nearly every other planet in the solar system spins like a top. Uranus is tilted on its axis almost 98 degrees. Like an enormous gyroscope, it rotates around the Sun. Part of the time, one pole of the planet points almost straight at the Sun. Uranus rotates on its axis once every 17.2 hours. But a year on Uranus is very long. It takes Uranus 84 Earth years to revolve around the Sun.

Voyager 2 traveled nine years from Earth for a look at Uranus, which told scientists a lot. Until then, they did not know how long it took Uranus to complete one rotation. Uranus, like Jupiter and Saturn, is a gas giant. It is made mostly of gases and ice. But unlike Jupiter and Saturn, Uranus probably has no rocky core, but a molten mixture of rock and ice fills most of the planet.

The upper atmosphere of Uranus is made mainly of hydrogen and helium. Below that layer, clouds of the gas methane wrap the planet. Methane absorbs red light, so Uranus looks green and blue from space. Winds create barely visible bands of clouds.

Astronomers are still studying the photographs Voyager sent back to Earth. They believe that, during the formation of Uranus, many comets collided with and became a part of the planet. When Voyager flew by the planet in January 1986, scientists were a little disappointed. The spacecraft could not see many individual clouds in the planet's atmosphere. Uranus still looked like a boring blue-green globe.

The most exciting pictures Voyager took were of Miranda, the smallest of the five major moons of Uranus. Miranda is a very strange moon. It has an icy surface marked with grooves, cliffs, and plains. Voyager sent back pictures of a canyon ten times deeper than our Grand Canyon.

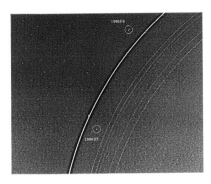

Shepherd moons

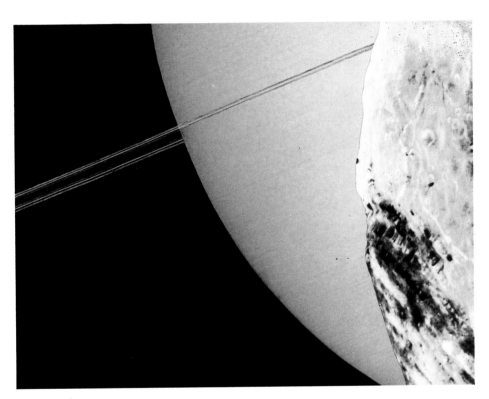

Until 1985, when Voyager neared Uranus, astronomers thought the planet had five moons. Voyager discovered ten more! Scientists think that some of these natural satellites are shepherd moons that help keep the rings in place. Voyager also showed more about the five middle-size moons that orbit Uranus. The two largest moons of Uranus are Titania and Oberon. They were named after characters in a play by Shakespeare. Oberon is much smaller than our moon. Voyager sent back pictures that show Oberon is very cold and has a rough, rumpled surface.

In the 1990s, six more moons were found orbiting Uranus. Nevertheless, we still don't know very much about this blue-green globe.

Above: Voyager 2 circles to see Uranus behind its small, strange moon Miranda. Left: This is an artist's drawing of what Voyager 2 looked like as it photographed Uranus.

NEPTUNE

Astronomers need a strong telescope to see Neptune. Even with a telescope, the planet looks like a fuzzy pale blue circle. Neptune is 30 times farther from the Sun than Earth is.

Neptune, the eighth planet from the Sun, looks a lot like Uranus. For years, scientists thought Neptune was the twin of Uranus. From Earth, Neptune is about the same color and size as Uranus. Neptune looks like a blue ball. Neptune is the smallest and the last of the gas giants. Fifty-eight Earths would fit inside Neptune. Scientists think that the inside of Neptune, like Uranus, is probably a molten mixture of rock and ice.

Astronomers studying Uranus wondered why its orbit was funny. They began to think another planet was pulling Uranus away from the Sun—they were right. They found Neptune.

Neptune revolves around the Sun once every 165 Earth years. Neptune has not yet completed a full orbit since it was discovered. Meanwhile, a day on Neptune is 16.1 hours long. Through a telescope, scientists found that Neptune, like the other gas giants, has a deep atmosphere that is mostly hydrogen.

Neptune was thought to have partial rings. Scientists knew it had moons—Triton and Nereid. Nereid is tiny and had the most elongated orbit of any moon in the solar system. Triton has always interested scientists. Triton orbits backward. It is the only large moon in the solar system that we know has a retrograde orbit.

Scientists knew that Neptune, like Uranus, has small amounts of methane gas in its atmosphere. It is the methane gas that gives Neptune its blue color.

Triton

Neptune is named after the Roman god of the ocean because of its blue color. The symbol of the planet is the three-pointed spear, called a trident, which Neptune used.

In the summer of 1989, when Voyager 2 approached Neptune, scientists learned a lot more about the smallest gas giant. Pictures of Nereid were poor because of the moon's great distance. But Triton was spectacular. Its south pole is covered with nitrogen ice, and beneath the surface, water ice is rock hard. Temperatures on Triton drop to more than 390°F below zero. Its scarred surface shows signs of volcanic activity.

Studying the information and photographs from Voyager was special because it was a joint effort. Scientists from the Soviet Union joined NASA scientists at the Jet Propulsion Laboratory, and they looked at the photographs together. Voyager showed the astronomers faint arcs, like dark rainbows, around the planet. Voyager proved theses arcs were three rings around Neptune. Another planet with rings!

For years we thought Neptune had just two moons. Voyager found six more. Scientists are still learning about them.

Scientists all around the world watched as Voyager 2 sped past Neptune and traveled farther out into the solar system. They were happy because they had learned so much. But they were sad, too, because Voyager was finished with its studies of the planets and would fly far beyond their reach into outer space. Neptune was the last new world they would see close up.

The volcanoes on the surface of Triton (right) are ice volcanoes. When the nitrogen and methane ices warm up, they turn to slush. Scientists think that the slush erupts much as lava erupts from volcanoes on Earth.

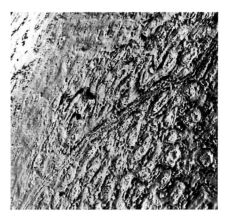

Neptune's rings

Another surprise Voyager discovered on Neptune is its violent weather. The planet is so far from the Sun that scientists did not expect much activity in the atmosphere. But Voyager 2 found winds swirling at speeds of 2,000 kilometers an hour. They named one storm system the "Great Dark Spot." It is a storm as big as Earth.

PLUTO

This artist's picture of Pluto shows how close its moon, Charon, is. Also, it shows how far from the Sun it really is—the Sun looks more like a bright star than the center of our solar system.

Because of its elliptical orbit, sometimes Pluto is the eighth planet from the Sun! Most of the time, it is the farthest, or ninth, planet from the Sun.

P Pluto was named for the god who ruled the underworld in old Greek and Roman legends. The name of the god—Pluto—is the symbol of the planet.

Pluto, the ninth planet in our solar system, is an icy ball. Even with a powerful telescope, Pluto is only a fuzzy pinpoint of light. The planet is about one fifth the size of Earth—smaller than our moon. Pluto is about 39 times farther from the Sun than Earth is. From Pluto, the Sun looks like a bright star that lights up Pluto's surface only a little better than a full moon lights up Earth.

Pluto is far from the Sun so its orbit is long. It takes the planet 248 Earth years to complete its revolution. And Pluto's orbit is strange. The path of its orbit is a stretched-out oval. If you hold a rubber band between the thumb and pointing finger of your hand and pull on the rubber band a little bit, it stretches out. That shape is similar to the shape of Pluto's orbit. It is the most oval, or elliptical, of any planet in the solar system. Sometimes Pluto is more than a billion miles closer to the Sun than other times. For 20 years out of every 248, Pluto moves closer to the Sun than Neptune. From 1979 to 1999, Pluto was orbiting at its closest to the Sun. During that time, Neptune was the farthest planet. It takes Pluto 6.4 Earth days to complete one rotation.

Pluto is so small and so far from Earth that scientists did not discover it until about 70 years ago. In 1905, an American astronomer named Percival Lowell thought the gravity of a planet was pulling on Uranus and

Neptune, making their orbits oval. Lowell predicted where the planet would be, but he died before he found it. In 1930, Clyde Tombaugh discovered Pluto after searching the sky for many years.

Pluto has one moon—Charon. Charon orbits so close to Pluto that scientists did not discover it until 1978. Astronomers found Charon because they noticed a bulge on one side of the planet in a photograph. That bulge was Charon. From Earth, the two objects look like they are touching. Charon is 20 times closer to Pluto than our Moon is to Earth. And Charon is about half the size of Pluto, making Pluto and Charon almost a double planet.

How the planet formed is still a mystery. It seems to be a snowball of methane gas, frozen water, and rock. Pluto probably has a surface of methane or nitrogen ice and a thin atmosphere of methane or nitrogen. It is tipped on its side like Uranus.

Someday we may find out more about the dark and distant world of Pluto. So far, no spacecraft has gotten near enough to the planet to answer our questions. But if you remember that we discovered Pluto so recently, you must admit we have learned a lot about it in a very short time.

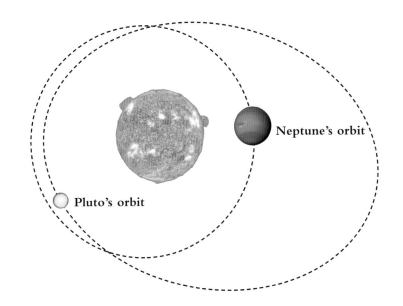

Neptune's orbit

Pluto's orbit

Pluto is dark and cold—temperatures drop to more than 300 degrees below zero. Life is impossible on Pluto.

Pluto and Charon from Earth (left) taken from the Canada–France–Hawaii Telescope. The picture on the right was taken from the Hubble Telescope.

 Pluto

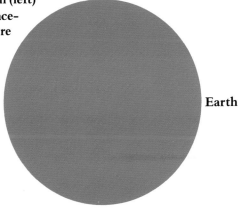 **Earth**

Pluto is about five times smaller than Earth, and is even smaller than our Moon.

Beyond us,

there is so much to explore. Earth is just a small planet that

revolves around an average-size star in a galaxy that has hundreds of billions of stars.

Most of our understanding of the solar system has come recently. When we launched scientific instruments into space, we began learning rapidly. NASA plans more  shuttle flights. The Hubble Telescope is sending photographs to Earth from space. A radio telescope, called the Very Large Array Telescope (or VLA), will help astronomers make maps of the universe. The world's largest telescope—the Keck Telescope in Hawaii—is taking pictures of new galaxies.

Each year brings new discoveries about our universe. It is so big that it contains endless surprises. We really have just begun learning about it.

42

GLOSSARY

Asteroid (AS-ter-oid): Small bodies, some less than half a mile wide to as much as 630 miles wide, that revolve around the Sun. Most of them are in the asteroid belt between Mars and Jupiter.

Atmosphere (AT-muh-sfiuhr): The gaseous outer layer of a planet.

Axis (AK-suhs): The center of a planet from the south pole to the north pole.

Comet (KAHM-uht): An object in the sky that usually has a large head and tail of glowing gas and shining dust when it is close enough to the Sun to warm up a little.

Constellation (kahnt-stuh-LAY-shuhn): Groups of stars that were named for a shape they were thought to have formed, such as Cancer the crab. There are 88 constellations.

Fusion (FYOO-zhuhn): A nuclear process that produces energy to make stars shine.

Galaxy (GAL-uhk-see): A large grouping of solar systems, stars, nebulae, and interstellar space. There are billions of galaxies in our universe.

Globular Cluster (GLAHB-yuh-luhr KLUHS-tuhr): A round collection of stars that looks like a dim, fuzzy star until magnified. A globular cluster may contain millions of stars.

Gravity (GRAV-uht-ee): The attractive force between any objects. Gravity makes objects fall toward the Earth and keeps the Earth revolving around the Sun.

Hemisphere (HEM-uh-sfiuhr): Half of a sphere. Earth is divided at the equator into the northern and southern hemispheres.

Interstellar Space (INT-uhr-STEL-uhr SPAYS): The area between stars.

Light-year (LYT-yiuhr): The distance light travels in one year, or about 5,878,000,000,000 miles.

Meteor (MEET-ee-uhr): A piece of comet or asteroid that lights up in the sky because of friction with Earth's atmosphere. A meteorite is a part of a meteor that falls to Earth.

Nebula (NEB-yuh-luh): A huge cloud of dust and gas in interstellar space that can be seen in the light of the stars nearby. A nebula can be the birth place of a star and what is left after a star dies.

Nova (NO-vuh): A sudden brightening of a small star, caused by rapid changes. A supernova is the rare explosion of a particular kind of star, where all the star's matter blows away and forms a nebula.

Orbit (AHR-buht): The path of a planet or other object around a larger object, such as the Sun.

Planet (PLAN-uht): A body in orbit around a star that is not large enough to glow with its own light.

Polaris (puh-LAR-uhs): The North Star.

Revolution (rev-uh-LOO-shuhn): A planet moving along its path or orbit. One complete revolution is a planetary year.

Rotation (ro-TAH-shuhn): A planet turning on its axis is its rotation.

Solar System (SO-luhr SIS-tuhm): A star, such as our Sun, and the planets, asteroids, moons, and comets that are held by gravity around the star.

Star (STAHR): A ball of gas visible in the night sky that glows by its own light. Our Sun is a medium-size star.

Universe (YOO-nuh-vuhrs): All the stars, galaxies, solar systems, and interstellar space.